Local Queens are Best - How to simply rear your own
B J Henderson Smith

ISBN: 978-1-914934-68-1
Text and photos © B J Henderson Smith except:
Fig. 12, Victoria de Labat; Fig. 24, David Evans;
Fig. 27, David Cushman / Roger Patterson.

Published by Northern Bee Books 2023
Northern Bee Books, Scout Bottom Farm, Mytholmroyd,
Hebden Bridge HX7 5JS (UK).
www.northernbeebooks.co.uk
+44 (0) 1422 882751

Book design by www.SiPat.co.uk

Local Queens are Best

How to simply rear your own

Bruce Henderson Smith

Contents

Preface

Say "queen rearing" to most beekeepers, and they will probably run a mile! This book aims to de-mystify the mysteries of simple queen rearing. It sets out to explain a method of queen rearing <u>on a small scale</u>, which can be used by any beekeeper with a few years of experience and a small number of hives, perhaps only three or four (or even two).

Fig. 1   A small apiary with 4 hives and a nuc box

Why raise queens at all, when the bees are perfectly capable of doing it all by themselves? Or they can be bought in?

I suggest there are three main reasons. The first is to improve the qualities of the bees we each have, such as good temper, hardiness, ease of management, and productivity. The second is to increase the number of colonies which consist of locally adapted bees. The third is that we can at once stop importing queens from abroad, and soon there will be no need to import queens from out of the county in which you live. The bonus is the satisfaction you will get when you see the first local queen you have raised heading her own colony!

Fig. 2   **A queen showing very visible signs of imported genes (Italian)**

Beekeeping should be enjoyable and fun. All of us will agree that it is far more enjoyable to look after and handle a good tempered colony. It also helps if the bees fly on cooler or wetter days which are not perfect, and are attuned to collect local pollen and nectar whenever it is available. It is also said that locally bred queens can mate in the vicinity of the apiary, and so do not need to fly away to a drone congregation area, which reduces the risk of losing her on a mating flight. These are the sort of qualities which we should all aim to encourage and develop in the colonies of bees that we look after.

It is now accepted by many experienced beekeepers that having locally adapted bees is much better than constantly (perhaps even every year) having to import queens from abroad, where the weather conditions and available forage that her genes are used to are likely to be very different from those where you live.

It is not only common knowledge amongst informed beekeepers, but also supported by scientific research carried out by members of the international honey bee research association COLOSS. A total of 621 colonies of 16 different genetic origins were set up in 21 apiaries in 11 different European countries managed by 15 research partners. The colonies were set up in the summer of 2009 and the results were published in 2014. The results showed that locally adapted strains of honey bee consistently performed better than "foreign" strains. Growing evidence was reported of the adverse effects of the global trade in honey bees, leading to the spread of novel pests and diseases. *(References and links are in Appendix 2).*

There is a real risk of importing diseases or pests with foreign bees or foreign queens, and they require strict inspection regimes to prevent this happening.

There is also the likelihood that the qualities for which the queens were imported will be lost within a few generations and the expense wasted. Any increase in genetic diversity is not universally beneficial, may contribute to colony losses, and is unsustainable in the long term. Personally I do not approve of hobbyists importing bees into this country.

Fig. 3   A queen showing obvious signs of DWV (deformed wing virus)

I must right at the beginning acknowledge the inspiration and help I have received from those in Cornwall who are already engaged in improving the stock of Cornish bees and raising black queens for that purpose. I particularly wish to mention Jo Widdecombe, whom I have known for many years. His book *"The Principles of Bee Improvement"* (2015) has been of great help and increased my beekeeping knowledge. It contains far more information and discussion of the principles involved than I can include in this short book. Reference should be made to it as often as the reader requires more detail.

Part 1: Introduction

The Beekeeper

I have been keeping bees in Cornwall for over 35 years. That means I started BV - before varroa. For most of that time, I was what I call a "weekend beekeeper," needing to allow for work and family responsibilities. Any time left over had to be shared between going to the beach and beekeeping. I kept no more than 6 Langstroth hives at any one time. I had not studied theory, or indeed practice, very much, and most of my beekeeping was at a simple level, that is, what was needed to keep my colonies going.

After retirement I had time to study beekeeping theory and practice in more detail, and was encouraged to study some of the BBKA modules. I probably learned more in 5 years than I had in the previous 30. However, I decided to keep less colonies, and was no longer interested in just producing masses of honey to sell. There was a trend to encourage more education, and queen rearing as well. I had learnt something about bee genetics in a study group for module 7 (the old version), and I decided I would give raising queens a go.

The Bees

During my weekend beekeeping, I had managed to keep a colony of bees in the same place in my garden for most of the 35 years. Through the occasional swarm, and natural selection, I discovered that I now had a colony of lovely, good tempered bees, which were well behaved, quiet on the comb and yet produced a good quantity of honey nearly every year.

This seemed to be the ideal colony from which to start rearing local queens. That would enable me to pass on the beneficial qualities of these bees to some of those in our local association group who were just beginning their beekeeping.

Objectives

Over the years I have formed strong views against the importation of bees into this country. Alongside these views, I have developed a belief that local Cornish bees are best suited to our very individual Cornish climate, and I support those who are trying to improve the qualities of local bees, especially good temper. I am a member of BIBBA and agree with the Bee Improvement Programme for Cornwall, both of which concentrate on breeding bees which are as black as possible without imported genetics or colouring.

Fig. 4   Seven supers on this Langstroth hive in August 2018

Holding those views, I wanted first to increase the number of colonies locally of good quality black Cornish bees.

This would avoid the need to import queens from abroad, or those which had been reared in other parts of the country, often from imported queens. Increasing the number of colonies in

Fig. 5   A local Cornish queen and her lovely black
                Cornish bees

Fig. 6   A more mixed-race queen and her offspring with some
                orangey-brown bands

this way would also help to increase the number of Cornish drones locally, which in turn would increase the chances of breeding mostly black bees, rather than unreliable multi-coloured mixtures.

My second objective was to be able to provide a nucleus colony of good quality, black Cornish bees to a few novice beekeepers in my group at a very reasonable price.

In achieving these objectives, I had every intention of working with the bees, not against them, and letting them do as much as possible in their natural way, rather than forcing my way of working on to them. I would also be able to develop my ability to *"think like a bee"*, which I believe is one of the most important attributes of a good practical beekeeper.

Fig. 7   Classic good quality sealed brood pattern with pollen and honey above

I freely admit that I am very keen on breeding black bees. It must be said that it is far from easy to obtain reliable and consistent results trying to do this. However, my feelings on this are not really the point. It's just how I started. The point is to try and improve your own bees, and those of other beekeepers near you, so that you all end up with better and calmer bees to handle in your colonies, well adapted and adjusted to your locality.

It's not just in our mild, wet and windy Cornwall. The principle is the same <u>wherever you live</u>. Bees reared in and used to your local climate, weather and forage will perform better than those bred by 'foreign' queens. There is just no need whatsoever to buy queens from outside your county. Good natured, calm, local queens, with offspring showing many desirable qualities, are well worth having, even if their colouring varies and is not wholly black.

Part 2: A bit of theory

Life cycle of the Honey Bee

I am trying to keep this book as simple as possible, but it is really essential that you should be familiar with the life cycle of the honey bee, and in particular of course the queen. You will find a chart on page 16. One word of warning. Many beekeepers work on the basis that the egg is laid on day "0", with the result that the queen emerges on day 15. However, I prefer to follow Mark Winston in his excellent book *"The Biology of the Honey Bee"* (1987). He has a clear chart on pages 50/51 showing the egg as laid on day 1, and the queen emerging on the traditional day 16. This also means that it is easy to remember that the larva is sealed in its cell on day 8 or 9 or 10 for queen, worker, and drone respectively.

Life cycle of the Honey Bee

Day number from laying of egg	Queen	Worker	Drone
Egg laid -- vertical in cell	1	1	1
Egg sags over to horizontal			
Egg hatches into larva	4	4	4
C-shaped larva grows rapidly			
Cell sealed	8	9	10
Larva 'moults' into pupa			
Pupa's eyes become red	12	15	
Adult emerges from cell	16	21	24
Mature and ready to mate	20	n/a	34
Workers as House Bees then foragers (and can sting!)		next 3 weeks	
Average lifespan as Adult	3-4 years (up to 5)	6 weeks (S) 4-6 months (W)	varies 3-5 weeks can be 2x that (3 months)

NOTE   These are the recognised average times, but can vary (eg with temperature) Queens can emerge a day early, or 1 or 2 days late.

Fig. 8   **Life cycle of the Honey Bee**

Queen Cells

The difference between a worker bee and a queen lies in the way it is fed. The egg is the same to start with. To develop into a queen, the larva is fed an abundance of royal jelly.

Fig. 9   Opened queen cells showing a larva floating in royal jelly

You can see in this photograph the C-shaped larva lying in a pool of royal jelly. It grows rapidly.

Fig. 10   Open extended queen cell with a large larva, nearly
           ready to be sealed

Fig. 11   An opened queen cell showing a developing queen
           pupa head downwards

When the cell is sealed, it turns into a large white pupa. In the hive, the queen cell hangs downwards from the frame. To start with, there is empty space below the young pupa. This is why you must not shake a frame with sealed queen cells on it, because you will dislodge the pupa from the royal jelly and it will fail to develop properly. The pupa grows head downwards, so that the queen is ready to bite her way out of the cell in order to emerge.

You will know there are three types of queen cell: for supersedure, for the swarming impulse, and emergency queen cells.

Fig. 12   A sealed supersedure queen cell, early in the season

Fig. 13   Two sealed swarm queen cells in the bottom corner of a brood frame

Fig. 14   Five emergency queen cells built on the edge of the area laid with eggs

Supersedure cells should not be used by the beekeeper for queen rearing, but should be left for the bees, who know what they are doing and have everything well under control.

Swarming queen cells can be used, although the genes involved may include a tendency to swarm, which you do not want to encourage.

My method uses the emergency cell impulse, as is used by many queen rearing methods. Some beekeepers say that queens reared under this impulse are not reliable or of good quality. I do not agree with that. There are many experienced beekeepers who happily use queens reared in this way, and my limited experience firmly supports this, providing a little care is taken, as is always needed in beekeeping procedures.

Fig. 15   Three quality sealed emergency queen cells

A quick word about the <u>physical appearance</u> of "good" and "poor" queen cells, and how to tell the difference between them. The first point to make is that the bees always know which is a good cell and which is a poor one. But the beekeeper often <u>does not</u>! It is said that a good cell has a broad base, is a good, mostly straight length, tapering symmetrically to a blunt end, and overall has an even crinkled appearance – whereas a poor cell is short, dumpy, oddly shaped or bent, and unnaturally crinkled.

Sometimes true, but often not! Many a time I have left cells of both types, betting that the long straight one will produce the best queen. When they have emerged, I find the opened cap on the irregular, bent cell, and the long straight one has been torn down. The queen that comes out goes on to do well. I agree that even crinkling is a good sign, as is the fact that the cell is being paid lots of attention with workers crawling all over it. Apart from that, try to leave it to the bees, who "always know best".

"Good" queen cells
Figs. 16–20

Fig. 16   **A good quality dimpled emergency queen cell, near the top of the frame**

Fig. 17   Two good quality sealed queen cells, though shorter and not so dimpled

Fig. 18   Similar good quality sealed queen cell, built with new pale wax

Fig. 19   **Good quality dimpled queen cell, after queen has emerged, showing cap**

Fig. 20   **Two reasonable queen cells, from which queens have emerged**

"Poor" queen cells Figs. 21–23

Fig. 21   Unsealed emergency queen cell with no dimples

Fig. 22   Poor quality queen cells, two joined together, one at bottom damaged

Fig. 23   **Apparently poor quality queen cells, although a queen has emerged (cell on the left)**

Inbreeding and Diversity

Forgive me if I introduce a little bit more technical science. You may be worried about inbreeding and diversity if you rear your own local queens. I don't think you need to be too concerned. At the moment, there are plenty of different drones around in this country and the ones that your virgin queens mate with are not likely to be related to her. There is a very interesting beekeeping blog written by David Evans on his website *"theapiarist.org".* In a recent item, he referred to diversity and polyandry. That topic is discussed in more detail in his blog entitled *"Who's the daddy?"* which was published on 29th January 2020. That considered a research paper by Withrow and Tarpy (2018). The conclusion reached in the paper is that bees behave in a particular way when raising an emergency queen.. The bees prefer to choose an egg/larva which has been fertilised with sperm from a drone whose total amount of sperm in the spermatheca forms only <u>a very small proportion</u> of the overall amount of sperm available to the queen who laid the egg. They choose to rear emergency queens from some of the rarest patrilines in the colony. This may act as a mechanism to maintain diversity. On that basis, using the emergency impulse to raise local queens is unlikely to result in inbreeding and the production of diploid drones. *(References and links are in Appendix 2).*

Part 3: How to do it

There is one more important point which I must make at this stage. The majority of queen rearing methods involve putting an egg or larva, <u>already in a cell</u>, into a queen cell raising colony. That colony (or part of it) is by definition queenless, so that the nurse bees in that colony raise replacement queens using the cells provided by the beekeeper. The result of this process is a number of sealed queen cells, which then have to be taken out and individually put into another (probably small) queenless colony to wait for the queen to emerge and get mated.

Fig. 24   **Ten quality sealed queen cells built on a cell bar frame, ready for harvesting**

That was not what I wanted, because it involves an extra manipulation, and extra equipment to be used as the mating colonies. It was too complicated. I wanted something simpler. My target was, in a single stage, to produce an already mated queen, laying and heading her own colony, which colony was then immediately ready to be used by me or by the novice beekeeper I was going to pass it on to.

Equipment

There is no need for lots of mating nucs, Apideas, mating apiaries, grafting tools, cell punches, starter colonies, cell raising colonies, finisher colonies, Vince Cook circles or whatever. Less expense, less need for storage space for infrequently used equipment.

You will need a few empty nuc boxes (which can be borrowed from reliable, bee-disease-free fellow beekeepers), or brood boxes and dummy boards. If you want, you can divide a standard brood box into two separate chambers, with separate entrances.

Fig. 25   Brood box with home made divider to allow two separate nucleus colonies

Fig. 26   Polystyrene (painted) nuc box with feeder and variable entrance disc

This is mentioned by Ted Hooper on page 175 of his well-known book *"Guide to Bees & Honey" 5th edition.* (NB Two half-sized wood crown boards with feeder holes are better than a canvas quilt). A polystyrene nuc box is lighter than a wooden one, and is recommended (because of insulation) if you want to overwinter one of your new queens. At least one nuc box is a good investment for any beekeeper and can be put to so many different uses.

Queen Rearing Timetable

The most important thing you need is the queen rearing timetable which can be found as an Excel spreadsheet on the Dave Cushman website, accessed from the following web page: *http://www.dave-cushman.net/bee/queenrearingtimetable*

The spreadsheet has been helpfully tweaked by Roger Patterson, who now owns and runs the website. You will find it useful to read what he says about the various methods of queen rearing, because this will improve your general understanding of the principles involved.

A copy of the Miller 1 chart is shown opposite.

BIBBA queen rearing time table. For Miller Method.

Enter start date larvae put in cell raiser DD/MM/YY :				07/05/2019	
			Weekday :	Tue	
	Day	Date	Wday	Q stage	Description
1	-5	28/04/2019	Sun		
2	-4	29/04/2019	Mon		
3	-3	30/04/2019	Tue		
4	-2	01/05/2019	Wed		
5	-1	02/05/2019	Thu		Comb placed in breeder colony
6	0	03/05/2019	Fri	Eggs laid?	
7	1	04/05/2019	Sat		
8	2	05/05/2019	Sun		
9	3	06/05/2019	Mon		
10	4	07/05/2019	Tue	12-24 hrs. old	Prepared comb placed in cell raiser
11	5	08/05/2019	Wed		
12	6	09/05/2019	Thu		Check number of Q/Cs started (not essential)
13	7	10/05/2019	Fri		
14	8	11/05/2019	Sat	Cells sealed?	
15	9	12/05/2019	Sun		
16	10	13/05/2019	Mon		
17	11	14/05/2019	Tue		
18	12	15/05/2019	Wed		Check number of Q/Cs built
19	13	16/05/2019	Thu		Distribute Q/Cs today or tomorrow
20	14	17/05/2019	Fri		Distribute Q/Cs latest
21	15	18/05/2019	Sat	Q's emerge	Could emerge today or over next 2-3 days. See Note 6.
22	16	19/05/2019	Sun		
23	17	20/05/2019	Mon		Check to see if Q emerged and check wings
24	18	21/05/2019	Tue		
25	19	22/05/2019	Wed		Latest day for opening colony between 10am-6pm
26	20	23/05/2019	Thu		

Please read "Instructions" Notes below before using.

Notes 1 This is a simplified version of the original "Tom's Table". Modified for the Miller Method

2 Column "B" with green background is for "conventional timing" with the day the egg is laid as being "Day 0".

3 Bold blue letters on yellow background indicate days on which action is required by thebeekeeper.

4 Entering the start date in cell F2 will automatically change all the other dates.

5 Bold white letters on red background show the date entered in cell F2.

6 Queens can regularly emerge up to 4 days overdue.

7 It is assumed the user knows when to perform all other tasks. These can be obtained from other literature.

8 This worksheet is unprotected when you get it. Protect apart from cell F2 and use your own password.

Fig. 27 **Queen rearing timetable - Miller 1 version** *(credit: Roger Patterson/BIBBA)*

The basics of this chart are as follows.

The first column is simply a numerical list of the number of days displayed. The second, green column is the number of the day of the queen's life cycle. Note carefully that the table uses day "0" for the day on which the egg is laid, so the queen is shown as emerging on day 15. The third column will automatically give you the calendar date of the day referred to in column 2. The fourth column will automatically give you the day of the week of the date referred to in column 3. The fifth column is self-explanatory. The last column contains notes or a description of the action to be taken. The notes numbered 1-8 underneath the chart should be carefully read and understood.

The most significant date is the date you enter in cell F2 ("start date"). This is then automatically repeated in cell C19 (red), and other dates and days are changed to match.

I have adapted this chart for my own use and a copy is on page 12. An explanation of the changes and entries is also there, because it can usefully serve as your record of what has taken place and what you have done.

And of course you will need a great deal of Patience!

Selection

For my method, the process of selection of your breeder queen is relatively simple.

You can make it more complicated if you wish, by keeping very detailed records, such as those suggested on the National Bee Improvement Programme hive record card, with space for many desirable traits.

However, if you have only three or four colonies, you will know which are your best bees. Which colony is best behaved, quiet on the comb during inspections, reasonable honey producers, less prone to disease, less likely to swarm, and needing little or no smoke?

Personally, I also prefer a colony where the bees are mostly of native appearance, although a colony with a small proportion (say 15-20%) of its bees having one orangey-brown segment is acceptable to me. This proportion of bees does not seem to affect the good qualities of the rest of the colony (bees of that colour may themselves also have those good qualities); and the colour of the bees may vary during the laying season anyway, according to the drone which provided the stored sperm used by the queen for that particular egg.

Very simply, a honey bee of native appearance will have a dark abdomen. This is divided into several segments, which can be seen quite easily. Each segment usually has a light coloured stripe on it (called a *tomentum*). The native bee has a thin stripe, whereas non-native bees have a wider stripe, sometimes as wide as the dark part of the segment or even wider. [Don't forget that even native bees are, in one sense, all mongrels anyway! Bee genetics can be *very* complicated] (Fig. 28).

In addition, a non-native bee is indicated by an orange or yellow or brown band on the abdomen. There may be only one such coloured band, or many (Fig. 29).

"Thoracic hair colour is also a good indicator of bee type. The area on the upper surface of the thorax has black and brown hairs in the dark native honey bee. Carniolans, Italians and Buckfast are paler, usually with no black hairs. The 'halo' of pale hairs around the thorax is very noticeable in the eastern

Fig. 28   Cornish dark native honey bees (and a drone)

types of honey bee and less prominent in dark native honey bees" *(according to the Scottish Native Honey Bee Society, see Appendix 2).*

In the colonies which I now have, I am not looking for the producers of the most honey, but rather the most enjoyable bees to handle, and the ones that are best able to survive disease and our local summers and winters with the minimum intervention from the beekeeper.

Fig. 29   Non-native honey bees, and their queen, showing signs of imported genes

You simply choose your best hive to breed from, and arrange to produce new queens from it. You can continue doing that from year to year. To breed better bees, you need to choose your <u>best</u> queen each year, and you may need to kill your <u>worst</u> queen and re-queen that colony with one of your new queens. This should result in the continuous improvement of your colonies and make your beekeeping far more enjoyable and rewarding. You will also be in a position to help your fellow beekeepers to benefit from your better queens. They can join in the process and perhaps together you can form a local bee improvement and bee breeding group (*see the BIBBA website*).

Method

As I have said, I didn't want to have to buy a lot of new equipment and I didn't want 20 plus virgin queens. All I wanted was a few new colonies each year so that I could start some new beekeepers on their way. So I chose a method of working with the bees where they would be doing as much of the work as I was.

The method is to use a 4-frame nucleus to make a new mated queen and colony.

Fig. 30   Making up a 4-frame nucleus to rear a local queen

For this you need 3 things -- plenty of eggs, good food supplies, and plenty of young nurse bees.

Having chosen the colony and queen which you're going to use for breeding purposes, you may need to prepare some time before the day chosen to set up the nuc. It depends upon your assessment of the breeder colony. The ideal is to have plenty of space for the queen to lay eggs in. You also need plenty of young bees, so you may need to start feeding a thin syrup early on to stimulate the queen into laying more eggs. You will have to use your judgement and experience.

By keeping an eye on the breeder colony, and the weather, you can usually plan in advance for a reasonably fine day, convenient to you, when y.ou are going to set up your first nucleus box to rear a new queen. You also need to be reasonably sure that there are going to be enough drones about, when needed, for your lovely new queen to mate with – so not too early in the season.

I usually take my nucleus box to the apiary already containing a frame of sealed stores. This will provide food for emergency purposes. It may be a surplus frame which you have kept from last year, or a frame which you have taken out of one of your colonies this year to provide more space for that queen to lay In. Put this frame next to the wall of the nuc.

Next go through your breeder colony and if possible find the queen. It will help if she is already marked. You can put her temporarily in some sort of cage or container, to keep safe. This will make sure that she does not get transferred into the nucleus box and it's then easier and quicker going through all the frames in the breeder colony. She won't get squashed!

EGGS. Now you need to choose two frames from the colony you are breeding from, with plenty of cells with an egg in, and there will probably be larvae and sealed brood as well. If possible,

Fig. 31   **What you are looking for - eggs newly laid in cells (mostly standing up)**

some unsealed honey stores are very useful, since it is generally thought that bees making new queens prefer to use unsealed honey for feeding them. Plenty of eggs will give the nurse bees the maximum choice from which to raise new queens, and they will make their own selection from whatever they think is suitable and at the right age. These two frames go into the centre of the box, next to the frame of stores. *(see also Appendix 2 about the bees choosing how to raise emergency queens).*

It can happen that the day you have chosen to set up your nucleus turns out to be a day when most of the available cells in the frames in the brood box have already been laid in, and most of the eggs have already hatched into larvae. In this case you will find that there are not many cells containing freshly laid eggs. (Remember that the queen lays an egg standing up apparently vertically (to us) from the bottom of the cell. That egg gradually tilts over, so that by day 3 it is lying apparently flat on the bottom of the cell. The next day it will hatch into a larva).

Most of the time, there are plenty of freshly laid eggs, but don't panic if that is not the case. Simply take a third frame which has fresh eggs in it, again if possible with unsealed stores, and put it in the nuc as your last frame, next to the middle two. So the bees should still be able to find enough eggs to make at least three or four queen cells, which will be enough. If you only need two frames with plenty of fresh eggs, you may find the bees make 10 or 15 queen cells! We'll come to that in a minute.

If you do only need two frames of eggs, then you have a bit of discretion about the last frame. It can be mostly empty cells, for your new queen to lay in; or have more unsealed stores in it; or be full of sealed brood, which will help to provide a useful workforce when the queen is starting to lay eggs and help to establish a decent sized nucleus colony. Or it may be a mixture of all of those.

It is also helpful to have stored pollen in all of the three frames taken out of your breeder colony, because this will be needed for feeding the queens and the young bees. If there is insufficient pollen in the frames, and the possibility that the bees cannot gather enough fresh pollen from local forage, you will need to feed a pollen substitute.

Finally I put a correx dummy board on the 'open' side of the frames, and move them reasonably close, always allowing room if necessary for queen cells to be made. They are usually made on the face of the comb, and not on the sides or at the bottom like swarm cells. They will be made where there are suitable eggs (hatched or about to hatch).

BEES. Some young nurse bees will have come attached to the frames you have transferred, but you will need more than that. They will be needed, not only to look after any unsealed larvae and the sealed brood, but also most importantly to choose and look after and feed up the new queens that are going to be made. If there are not enough nurse bees, the process will not work. That has happened to me. And they need to be there now. THIS IS ESSENTIAL. It is also likely that some of the bees on the frames that you transferred will have been foragers, who may not stay in the nucleus, but fly off and return to the breeder colony.

You will have to use your judgement and will probably need the house bees off at least two more brood frames. Use the two-shake process, dislodging the foragers first. Possibly you cannot have too many house bees, although you do need to think of overcrowding, and overheating if it is going to be very hot and you have not given your nucleus colony some covering shade. At worst, the nuc might swarm -- yes, that has also happened to me! (Or totally abscond, as can happen with a mini-nuc like an Apidea if there are too many bees in it).

FOOD. It is also essential that the nurse bees have enough food at their disposal. It is quite likely that there will not actually be enough available in the frames you have transferred, and there will be few, if any, foragers available in the nuc to go out and fetch extra food in. It is therefore a good idea to always feed thin syrup straightaway when you set up the nucleus. As already mentioned, if pollen is in short supply, you need to feed a substitute or use some sort of supplement.

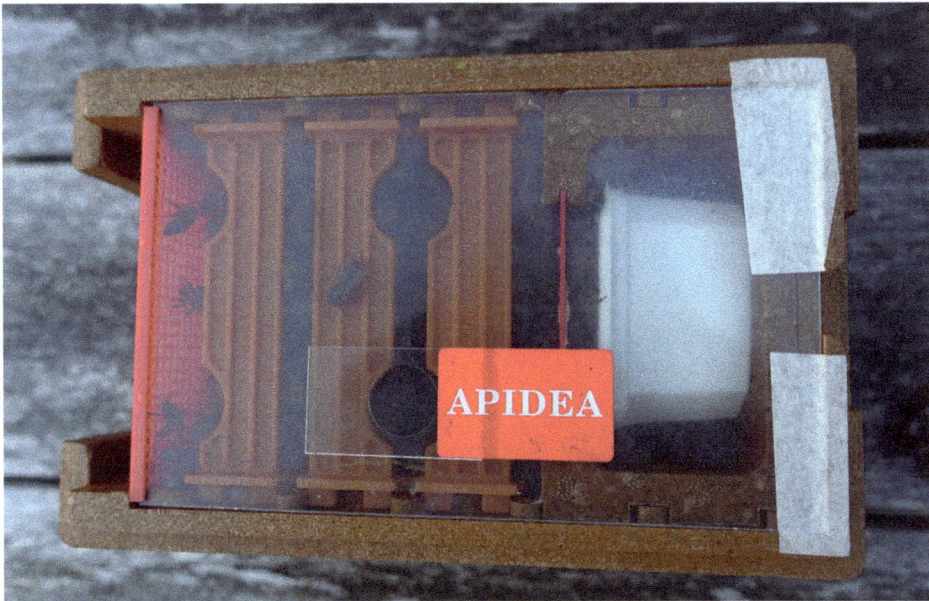

Fig. 32   An APIDEA mini-nuc (roof off) filled with too many bees, and candy for feed

I prefer to close the entrance to the nucleus colony for 24-48 hours, even if you are moving it to another apiary. Partly this is to help avoid any possibility of robbing, but it also allows the bees to adjust and recognise that they are a new colony. The bees in the nucleus will quickly realise that they are queenless and will start to make emergency queen cells.

BABYSITTING. Then it's over to the bees to do their part of the work! But I do like to keep a very close eye on them, and check them regularly, in line with the Cushman chart I have mentioned. The nucleus doesn't seem to mind being looked at every few days, as long as you are gentle, and I don't usually use any smoke. If there are masses of queen cells, I usually reduce the number, but leave between 5 and 8. There are always some blanks and the bees will choose the best ones themselves. We often disagree!

It is possible to correct the situation a few days later, if you are short of nurse bees, or short of food, and perhaps find that no queen cells have been started. More bees can be added from any of your colonies, even from super frames. At a pinch, if you really think that there were not enough fresh eggs, you could add, or exchange, another frame of eggs from your breeder colony and hope that the bees will start again, because they will still know that they are queenless.

It is also a good idea to check on day 4 and remove any queen cells which might by then have been already sealed on older larvae. This is because a queen from an older larva is not likely to have been fed or eaten enough royal jelly to develop fully. It's better to make sure that the bees only use younger larvae, which they can feed to the maximum in the usual way. *(Appendix 2 contains details of some research into how the bees decide which eggs or larvae to use – helping the beekeeper).*

ADAPTED CHART. An example of the queen rearing chart as I have adapted it is on the opposite page. This is one that I used towards the end of June 2022 and it has my notes on it of what happened.

You will see I actually started the nuc on 23rd June and I treated this as day 1 on the chart. This was because the eggs in the cells looked flatter, and I thought they had been laid on the day *before* I started the nuc (i.e. laid on the day marked 0 on the chart).

For my purposes I then had to work out what date to put in cell F2 as the 'pretend' start date when larvae would have been put in the cell raiser colony (Miller method), so that the chart would work for me and fit my method. With grafting etc, the

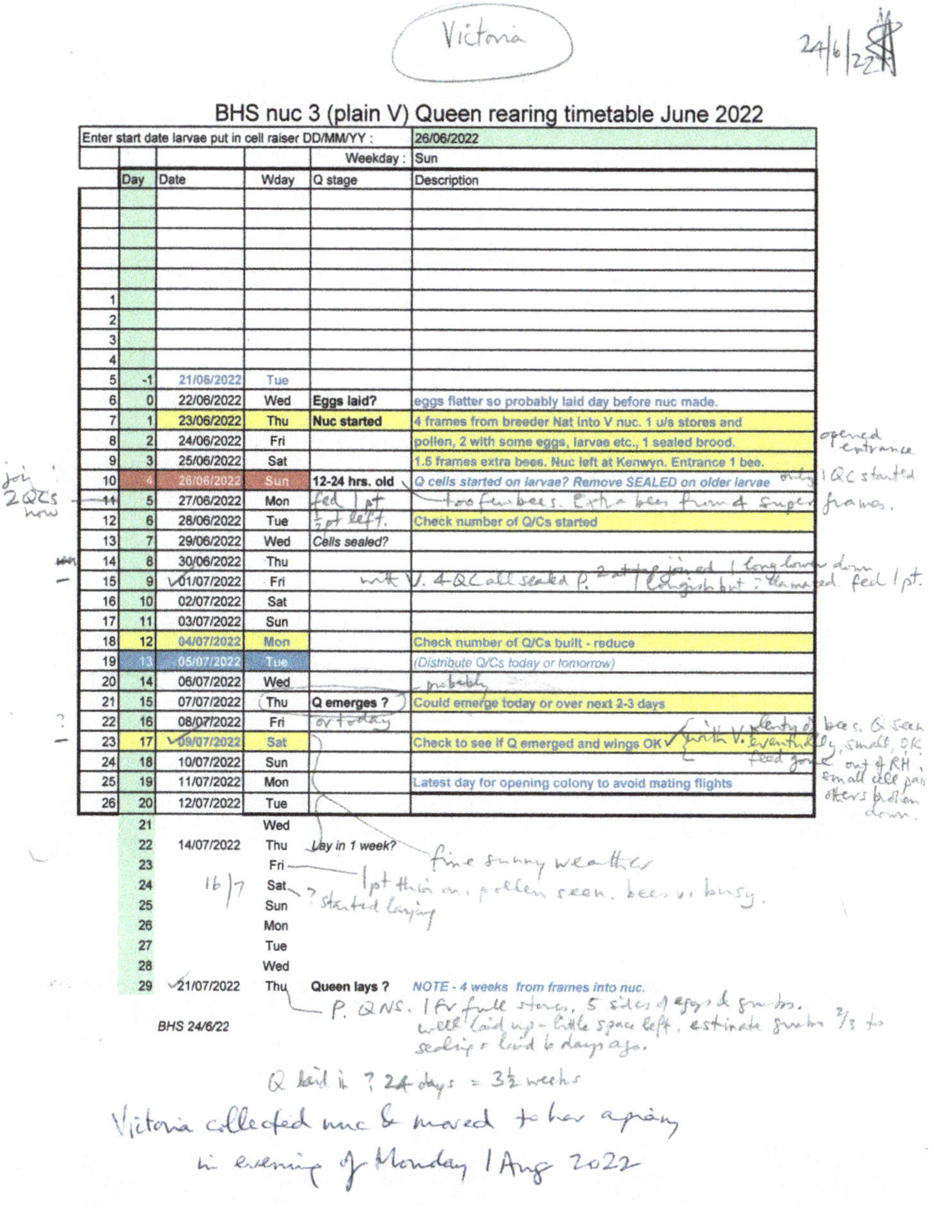

Victoria   24/6/22

BHS nuc 3 (plain V) Queen rearing timetable June 2022

Enter start date larvae put in cell raiser DD/MM/YY :				26/06/2022	
			Weekday :	Sun	
	Day	Date	Wday	Q stage	Description
1					
2					
3					
4					
5	-1	21/06/2022	Tue		
6	0	22/06/2022	Wed	Eggs laid?	eggs flatter so probably laid day before nuc made.
7	1	23/06/2022	Thu	Nuc started	4 frames from breeder Nat into V nuc. 1 u/s stores and
8	2	24/06/2022	Fri		pollen, 2 with some eggs, larvae etc., 1 sealed brood.
9	3	25/06/2022	Sat		1.5 frames extra bees. Nuc left at Kenwyn. Entrance 1 bee.
10	4	26/06/2022	Sun	12-24 hrs. old	Q cells started on larvae? Remove SEALED on older larvae
11	5	27/06/2022	Mon		
12	6	28/06/2022	Tue		Check number of Q/Cs started
13	7	29/06/2022	Wed	Cells sealed?	
14	8	30/06/2022	Thu		
15	9	01/07/2022	Fri		
16	10	02/07/2022	Sat		
17	11	03/07/2022	Sun		
18	12	04/07/2022	Mon		Check number of Q/Cs built - reduce
19	13	05/07/2022	Tue		(Distribute Q/Cs today or tomorrow)
20	14	06/07/2022	Wed		
21	15	07/07/2022	Thu	Q emerges ?	Could emerge today or over next 2-3 days
22	16	08/07/2022	Fri		
23	17	09/07/2022	Sat		Check to see if Q emerged and wings OK
24	18	10/07/2022	Sun		
25	19	11/07/2022	Mon		Latest day for opening colony to avoid mating flights
26	20	12/07/2022	Tue		
21			Wed		
22		14/07/2022	Thu	Lay in 1 week?	
23			Fri		
24		16/7	Sat		
25			Sun		
26			Mon		
27			Tue		
28			Wed		
29		21/07/2022	Thu	Queen lays ?	NOTE - 4 weeks from frames into nuc.

BHS 24/6/22

Handwritten annotations:
- join 2 QCs now
- opened entrance
- only 1 QC started
- (row 27/06) fed pt — too few bees. Extra bees from 4 super frames.
- (row 28/06) ½ pt left.
- (row 01/07) with V. 4 QC all sealed P. 2 others joined, 1 long low down, longish but thumbnailed feel 1 pt.
- probably
- or today
- with V. eventually, plenty of bees. & seen small, OK feed gone out of RH, small cell pan, others bottom down.
- fine sunny weather
- 1 pt this on. pollen seen. bees v. busy.
- ? started laying
- P. QNS. 1 fr full stores, 5 sides of eggs & grubs. well laid up - little space left. estimate grubs 2/3 to sealing + laid 6 days ago.
- Q laid in ? 24 days = 3½ weeks
- Victoria collected nuc & moved to her apiary in evening of Monday 1 Aug 2022

Fig. 33 The Miller 1 queen rearing timetable, adapted by the Author, in use in 2022

43

beekeeper chooses larvae which are 12-24 hours old, which is when the bees start feeding a larva extra food for her to become a queen. That is day 4 on the chart, counting the egg as laid on day 0. Since my conclusion was that the (oldest) eggs could not have been laid earlier than 22nd June, the date to insert in the spreadsheet was 26th June. Of course some eggs could have been laid on the day I started the nuc, but with my method there will always be some variation about the precise day on which an egg was laid. Don't worry – leave it to the bees and they will sort it out!

On day 4, I found that only one queen cell had been started and decided this was because there were too few bees. I therefore added extra bees from 4 super frames. This improved the situation. I didn't actually look at the nuc again until day 9, by which time there were four sealed queen cells. I had

Fig. 34   A virgin queen, crawling over workers who ignore her. Note size and colour of legs

started feeding quite early and fed again on day 9. There was no need to reduce the number of queen cells because there were only four. On day 17 we checked to see if the queen had emerged. We eventually saw her. She was then quite small, but acceptable, and obviously was a virgin queen and not yet mated.

The next several days were fine sunny weather and the colony was not opened during that time. This is standard practice when the queen might be taking her mating flights, to avoid any disturbance to the colony and any chance of damaging the queen, or her becoming confused and not returning back to the correct nucleus colony.

This is a time for great patience, because it can easily take 2-3 weeks from emerging for the new queen to mate and come in to lay, for all sorts of reasons, but especially if there should be any bad weather.

On day 23, I saw that the bees were very busy and taking pollen in and fed one pint of thin syrup. I did not actually look in the nuc again until day 29. I did not see the queen, which is not unusual for one which has only just started laying, but there was good news! There were five sides of comb (that is two and a half frames) with eggs and larvae and it was all well laid up in a good pattern with little space left. I estimated that the oldest larvae had grown to two thirds of the way to being sealed and therefore had been laid, say, six days previously. Counting back would mean that the queen might have started laying on day 24, Saturday 16th July. That was consistent with my seeing pollen going in on day 23. If correct, that would mean that it was only three and a half weeks between the egg being laid and the queen starting to lay. This was a very satisfactory outcome, and helped by some fine weather.

This nucleus was destined for a fellow beekeeper and she collected it on 1st August. The colony established itself in a full-sized brood box and survived the following winter into 2023.

It is possible to take another 3 frames out of your breeder colony a few days, or a week, later and set up a second nucleus. Obviously you will need to replace the frames taken out, preferably with drawn comb. There will be 5 original frames left behind with the laying queen, which should allow the colony to recover from what you have taken out. You might also be able to repeat the process and take a third nucleus out later in the season.

Please try to have a go at rearing your own local queens. It's great fun! You don't need perfect weather. Just some organisation, some concentration, and a good deal of patience – and of course friendly bees!

I hope you can see the method to be followed. I have mentioned some of the difficulties that you can come across. However, my results have mostly been good. If yours are too, you will get a decent mated queen, who will start laying delightful local bees. You will have a new colony and a new *local queen*, which you have reared yourself (but with a little help from the bees!).

Good luck!

Appendix 1: Results and more photographs

2019

This was my first year. I tried the Miller method, which was recommended to be one of the simplest. I used a super frame.

Fig. 35   Frame of eggs, with wax cut to Miller shape, being put into the queenless hive

I was only working with one colony of bees, so I had to temporarily remove the queen to a nucleus box, so that the colony was queenless and would make queen cells on the frame. They worked hard and produced an abundance of cells which you can see in the photograph.

Fig. 36   The bees have busily made all those queen cells!

A couple of days before the queens were due to emerge, we cut out the best cells and distributed them amongst some fellow beekeepers.

I think the introduction of the cells was not straightforward, but I have no detailed records.

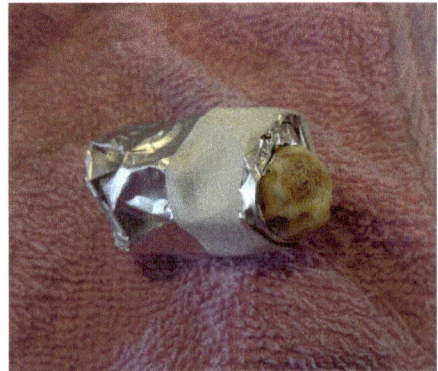

Fig. 37   Cut out queen cell, with protection, ready to be inserted into queenless nuc

When I tried to re-introduce the stored queen back into her own hive, the bees did not like this and unfortunately killed her. They preferred to raise a new queen of their own, in the

Fig. 38   One way of inserting a queen cell into a nuc

colony, from eggs which she had laid. So I lost the breeder queen, but gained a new queen from the same genetic line.

As mentioned, this method did not appeal to me, so I decided to try something different.

2020

For this year I converted a Langstroth brood box into a double nuc box as mentioned by Ted Hooper. I made a plywood divider down the centre, which reduced the available space to two 4-frame nucs, rather than the original 10 Hoffman frames which fit into a Langstroth. I made a simple plywood floor which incorporated a small entrance for each nuc at opposite ends of the brood box. I painted a blue surround on one entrance, and yellow surround on the other, to help the bees and the new queens to identify which colony they belonged to.

Fig. 39   Converted Langstroth brood box showing different coloured entrances

Fig. 40   One 4-frame nuc in the converted brood box. Oilcloth cover fixed centrally

Fig. 41   The other 4-frame nuc. Both have dummy boards

I successfully produced two colonies headed up by new queens, but still had the physical problem of transferring the colony from Langstroth frames into a National brood box.

Most beekeepers in Cornwall use Nationals. Whilst the transfer *can* be made (using a homemade conversion board inserted between the two boxes), it is rather a tedious and long-winded process, if significant losses of bees are to be avoided.

2021

This year I decided to move my chosen breeder queen into a National brood box to start with. I would then be able to take National frames out to put into National nucs and thus create a new colony with a new queen all already on National frames, which could easily be transferred into the recipient's brood box.

Unfortunately, the bees decided to supersede the queen in the hive that I had selected to breed from. This was in the spring. So I had to choose a different queen, and get her into the National brood box pretty promptly, so that that colony had time to develop sufficiently... but that's another story!

Fig. 42   An impromptu tower being used to move the queen into a National brood box!

I aimed to produce 3 new nucleus colonies. One failed completely, apart from producing a swarm with a virgin queen. Fortunately they only went into a friendly neighbour's garden.

The swarm was caught and passed on to a beginner. The queen mated successfully and it became a full colony.

Fig. 43   The small swarm with a virgin queen, which came from the nuc box

One nuc succeeded well and was passed on. In addition, whilst examining that nuc, I was able to release an emerging virgin queen from a spare queen cell on one of the frames, and pop her into a cage (rather old-fashioned and possibly home-made).

Fig. 44   Unusual, but very practical, queen cage

A fellow beekeeper I knew had a queenless hive and I was able to give her the virgin queen in the cage. The sealed cage was put in the queenless hive to allow familiarisation, and she was successfully released a few days later.

Things were now going so well that I then went a bit mad. I had intended to pass on the prolific breeder queen as the third nuc. Instead, I decided to take her out and put her in a nuc of her own. I was then able to overwinter that nucleus colony, which was then donated to our group apiary and has developed into a full-sized colony.

Fig. 45   **Feeding candy to the nuc after Christmas. They really needed it**

The then queenless breeder colony produced a new queen of their own, which I decided to keep and use for breeding the following year. They also made so many queen cells that I made up yet another nuc, with one of their frames that had five cells on it. That extra nuc successfully produced a mated queen, and the colony was used to help out a novice beekeeper who had a couple of colonies without viable queens in them.

I also made a late nuc for myself from my own favourite Langstroth hive in early July. The queen mated successfully and the colony became an extra hive in my other apiary.

Fig. 46   The small, but successful, 4-frame nuc

Fig. 47   That nuc with the (traditional) grass stuffed in the entrance to close it

Fig. 48   **Four frames with plenty of bees in the Langstroth nuc box**

So in 2021, I bred 6 queens from my good, calm Cornish bees, and I helped 4 beekeepers with local bees that should be a pleasure to inspect and learn from. In addition, my original breeder queen went to help in the group apiary, and I gained an extra colony. I was kept quite busy!

2022

By now, I had gained experience in raising these local queens by using my nucleus method. It was fairly straightforward to produce three more new colonies, using the breeder queen that I had kept in the National hive towards the end of last season. These three all went to help start off local beginner beekeepers.

Fig. 49   Laying workers. You can just see several eggs in many of the cells

Fig. 50   Laying workers. Sealed drone brood in worker cells, with domed caps

Unfortunately, my own beekeeping was not quite so successful. The extra Langstroth hive (which went into my other apiary in 2021) must have swarmed several times whilst I was busy doing other things with family visitors. The final replacement queen for some reason did not succeed, and the colony suffered robbing from some of the many other hives which are situated round and about where I live. The colony did not survive. (BeeBase shows that there are over 200 other apiaries within 10km of my apiaries).

A similar fate seems to have befallen the National hive from which I had earlier bred my three nucleus colonies. I had a very good year for honey, but the downside seems to have been multiple swarming, which I did not properly attend to. Again, the final replacement queen in the National did not succeed and I ended up with laying workers and the colony not surviving (Figs. 49 and 50).

Something of a setback, although I still had other good queens that I felt it would be worth breeding from the next year. I also thought I could ask those who have had a nucleus from me to let me have 3 frames out of one of their hives so that I could continue breeding with the same genetic line.

2023

I am writing this during the 2023 season.

Two out of my three hives joined in the national fever and swarmed themselves. The first swarm was on April 17th. I caught that one, but she swarmed again, and she and her crowd ended up on the ground.

Fig. 51   Collecting the swarm from the ground using an empty box with two frames in

Fig. 52   Overcrowded nuc transferred into its new home. Unfortunately they swarmed!

This time, they were all shaken into a National hive and went to a beekeeper in our group. So far as I know, they are still there.

The colony that swarmed first then threw a cast swarm, despite my reducing the number of queen cells. I caught that one too, and put it into a nuc box. The queen mated and energetically filled the nuc to the brim. This went to another beekeeper in our group.

The second prime swarm escaped. The new queen was lost somehow. Eventually a second frame of eggs put into the hive in July prompted the building of 4 emergency queen cells. I put in some extra nurse bees as well, which I am sure helped. One cell produced a queen, and the other three were not used. Of those, when I inspected them, one had been torn down by the bees and was empty, one (still sealed) had a queen about half way through pupating, and the last one had a pupa which was still white.

Fig. 53   All white queen pupa in her sealed cell which I opened to have a look

I hope the queen that emerged will mate successfully (despite the awful weather) and take the colony through the winter. I shall then be back to three hives.

[PS Success! That queen mated and has started laying eggs in a good pattern. My adapted chart records that she mated in about 2 weeks after emerging and started laying in about 2½ weeks. There were about 4 weeks between putting the frame of eggs into the hive, and her starting to lay. That is quite quick and she actually emerged earlier than I was expecting. I just hope that she found some burly Cornish drones and has mated properly.

PPS   Managed to save her and her bees from starvation by some emergency feeding in early September. Phew! Always bee on the lookout!].

Incidentally, the colony that swarmed first had had an unusual queen (for me) with many orangey-brown stripes (see Fig. 2). Her egg must have had a mixture of genes, probably including some from an imported Italian queen. She was producing workers with one or two distinct orangey bands, and was very prolific.

In contrast, the eventual replacement queen now leading that colony is beautifully black, and seems to be producing all black (native) workers. So perhaps I am nearly back to my original breeding line that I had in that hive before starting in 2019. (I mark all my queens white for simplicity. My records show the year of the queen).

So next year I can resume breeding from that colony, which were always my calmest and most good tempered bees. I am proud to say that members of my group who have had queens and nucs from me continue to talk about how calm "Bruce's bees" are.

The difference in these queens only goes to show how difficult it is to maintain a pure breeding line with honey bees. The

Fig. 54   My new Cornish queen and her black offspring

Fig. 55   Posing for a close up

behaviour and appearance of the offspring of any wonderful queen are dependent on the drones she mates with, in addition to her own genes. In most areas, we have little or no control over those drones. This can produce random results as described above. But this difficulty does not take away the pleasure of rearing your own local queens. A calm local queen is worth having, whatever her colour scheme.

Roll on 2024! Please note - you never stop learning about bees, learning from your own experiences, and constantly meeting the challenges the bees put in your way.

Learn to think like a bee!

Appendix 2: References, sources and acknowledgements

Page 1	<u>The COLOSS study</u>. Carried out from 2009 and results published in 2014. See the article in *"Entomology Today"* of July 15 2014 published by the Entomological Society of America at *https://entomologytoday.org/2014/07/15/are-local-honey-bees-healthier-than-imports* and a brief note about a special issue of the *"Journal of Apicultural Research"* dated April 15 2014 at *https://www.beebreeding.net/index.php/2014/04/15/journal-of-apicultural-research-jar-gei-special-issue-2014/*
Page 4	Photograph of supersedure queen cell taken by Victoria de Labat. Used with permission and with many thanks for her help throughout.
Page 5	*"Who's the daddy?"* An article posted on January 31 2020 (look in his archives) in the beekeeping blog by David Evans on his website at *https://theapiarist.org.* The blog is full of common sense and wisdom about all topics to do with keeping bees and looking after them. Regular references to the latest research, and thoroughly recommended.

Page 5	The research paper considered by David Evans was published by J M Withrow and D R Tarpy in PLOS ONE on July 11 2018 and is entitled *"Cryptic "royal" subfamilies in honey bee (Apis mellifera) colonies"* and can be accessed at *https://journals.plos.org/plosone/article?id=10.1371/journal.pone.0199124*
	Photograph of capped queen cells on a cell bar frame taken by David Evans and referred to in his article *"More queen rearing musings"* posted on June 17 2022. Used with permission and thanks (and a coffee or two).
Page 6 (and 11)	I could not have organised my attempts at queen rearing without the queen rearing timetables that appear on the Dave Cushman website: *http://www.dave-cushman.net/bee/queenrearingtimetable*. So I must acknowledge a large debt of gratitude to BIBBA; and to Roger Patterson who rewrote the tables in 2015. As he suggests, I have tweaked his version to suit the method I use. But his framework is so helpful that I even use it when just putting a frame of eggs into a queenless hive. He is absolutely right that the event dates are not set in stone but can vary considerably! That makes it all the more interesting!

Page 7	The Scottish Native Honey Bee Society. I don't pretend to be either a biologist or a scientist, and I have used what information I have found which I think will help the readers of this book to understand what I am saying. This paragraph is a quote from a webpage within the "Honey Bee Identification" section of the Society's website and the link is *https://www.snhbs.scot/scoring-hairs-and-tomenta-on-worker-honey-bees.* Obviously Scottish bees and English bees are not 'local' to the same place, but there is a lot of useful information on this website, expressed in simple terms. My thanks for the help.
Page 12 (and 10)	With regard to checking on day 4 and removing sealed queen cells, there is another interesting research paper by D R Tarpy *et al* entitled *"Honey bee colonies regulate queen reproductive traits by controlling which queens survive to adulthood"* at Insect. Soc. 63, 169–174 (2016). It is suggested that the bees choose to raise emergency queen cells from 3 day old eggs or 1 day old larvae and do not use queens started on younger eggs or older larvae. See also David Evans' article "The bees know best" posted on June 3 2022.

We all need somewhere to look up more details when we need help on how to do something or why something happened or what on earth the bees are playing at. There are many useful books, but my first "go to" is usually the Dave Cushman website, often followed by *theapiarist.org*. I also find the new *BBKA NEWS Special Issue Series* full of good information.

Finally, I would like to say a big 'thank you' to several beekeepers in the Cornwall BKA and the West Cornwall BKA, far more experienced than I am, who have prodded and prompted me both to improve my practical skills and to study for some of the modular exams. I would not be the beekeeper I am today without their encouragement lasting over several years. I am not going to name two in particular – they know very well who they are. My bees and I thank you both.

About the author

Bruce Henderson Smith lives in Cornwall and has been keeping bees simply for over 35 years.

His father used to keep bees near London before and during the Second World War, and Bruce grew up with the aroma of old bee equipment in their garage. It was actually his father-in-law, in Cornwall, who persuaded him to take up the hobby.

After retirement, Bruce had the time to learn more about better practical beekeeping and study some of the theory. In his 70s, he passed two BBKA modules, one with distinction and one with credit.

He has shown honey at the Royal Cornwall Show, also the National Honey Show, winning many cups and prize cards. He is an Honorary Life Member of the Cornwall Beekeepers Association and a past Chairman of the CBKA Council.

As a change from honey production, he started queen rearing in 2019. He is very keen on encouraging and persuading beekeepers of all standards to learn more. Hence this book to improve both local bees and the expertise of local beekeepers everywhere.

www.ingramcontent.com/pod-product-compliance
Lightning Source LLC
Chambersburg PA
CBHW061457270326
41931CB00021BA/3486